D1337618

www.thepig-club.com

This is a Carlton Book

Published in 2009 by Carlton Books Limited
20 Mortimer Street
London W1T 3JW

10 9 8 7 6 5 4 3 2 1

A CIP catalogue record for this book is available
from the British Library.

ISBN 978 1 84732 454 2

Publishing Manager: Penny Craig
Project Editor: Jennifer Barr
Managing Art Director: Lucy Coley
Design: Zoë Dissell
Production Controller: Luca Bazzoli

Printed in China

Artlist Collection
THE PIG ™

LITTLE **PIGGIES**

CARLTON
BOOKS

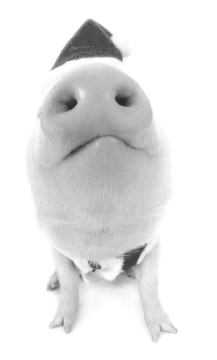